BRAINY SEX
VOLUME I

SCIENCE & LUST

Written by
REBECCA COFFEY

BRAINY SEX
VOLUME 1

SCIENCE AND LUST

Copyright © 2018 Rebecca Coffey
www.RebeccaCoffey.com

ISBN: 978-0-9972644-3-2

All rights reserved. No part of this publication may be reproduced, distributed, performed, or transmitted in any form or by any means without the prior written permission of the publisher, except in the case of brief quotations embodied in critical reviews and certain other uses permitted by copyright law. For inquiries, contact Rebecca Coffey, www.RebeccaCoffey.com.

Beck & Branch Publishers

CREDITS

Many of these essays were originally published on PsychologyToday.com as part of Rebecca Coffey's column, "The Bejeezus Out of Me."

An early version of the essay "The Erotic Psychoanalysis of Anna Freud by her Father, Sigmund" is also available as an ebook and paperback. The essay itself is the "truth of the matter" document that guided Coffey in writing *Hysterical: Anna Freud's Story* (She Writes Press, 2014). *Hysterical* is a fact-based, fictional memoir of Anna as she comes of age sexually and is psychoanalyzed by her father.

SCIENCE AND LUST

REBECCA COFFEY

BRAINY SEX
VOLUME 1

SCIENCE & LUST

CONTENTS

Do Pygmy Chimps Dream of Electric Lips?

Maybe you've already heard.

In the fall of 2016 a 1993 study about rats wearing polyester won one of Harvard University's Ig Noble prizes, awarded to *bona fide* scientific achievements that "first make people laugh, and then make them think."

Egyptian sexologist Ahmed Shafik had dressed 75 male rats in different types of pants. Some pants were polyester. Some were a polyester-cotton blend. Some were cotton. Some were wool. The rats had to wear the pants for 12 months. During the course of that year, the rats wearing polyester had far less sex than the other rats. Shafik speculated that the electrostatic potentials of their polyester pants generated electrostatic fields that reduced the sex drives of those rats.

And he may have been right. But something else might have accounted for at least some—and maybe even all—of the reduced sexual activity.

Could it be that no female rat of discernment wants to have

sex with a guy wearing polyester? Rather than showing an effect on sex drives of electrostatic fields, did the Shafik study reveal a cultural impediment to intimacy in rats? In rat society, is polyester passé?

And, if cultural differences can ruin a rat's intimate life, can they also lay waste to a human's?

Look First to Freud

In his 1930 treatise *Civilization and its Discontents*, Sigmund Freud claimed that humans need help keeping innate, beastly impulses at bay. He thought humans are more individualistic than communally minded, and he said that their selfish, instinctive urges toward pleasure create psychological trouble when they bump up against civilization's demand for conformity. Freud believed that men and women are hardwired to be as self-centered and destructive as the great apes are, and that it is civilization that forces them to live cooperatively.

That point of view pretty much prevailed until the late 1960s when British zoologist Desmond Morris suggested that Freud might be wrong about humans—and about all great apes. Morris thought that humans aren't self-centered and destructive by heritage because the great apes aren't self-centered and destructive. Like Freud, who was somewhat of a genius at catching headlines, Morris was a great front man for his own ideas. He wrote *The Naked Ape*, which in 1969 was published to much acclaim. In it and in interviews that were part of his book tour he made just enough surprising observations and speculations to—well, to sell 23 million copies. Here are two such observations and speculations.

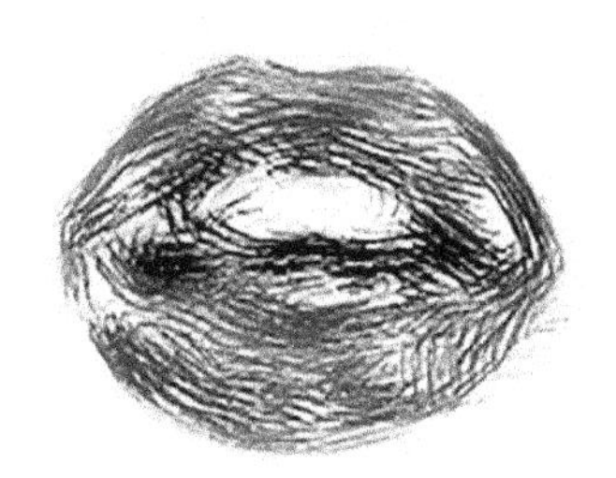

Observation: Women evolved larger breasts and plumper hips than are typical of non-human primate females. Speculation: Walking upright meant that their swollen, red labia were no longer "in men's face" so to speak. Breasts and buttocks evolved to serve as attractants.

Observation: Relative to body size, men have the biggest penises of all primates. Speculation: This, too, is a direct result of erect posture. Once everyone was standing up and frankly assessing each other, men, like women, needed an attractant.

Unlike Freud, who thought that humans' bestial heritage is a problem that needs reigning in, Morris proclaimed that humans' heritage helped both humans and their primate ancestors thrive. Yes, it may have been six million years ago that the human line separated from that of its primate cousins. But at least around 44 million years of primate hardwiring predated that split. Biologically and neurologically, modern humans are much more animal than human. We may be animal culturally, as well. Great apes today live cooperatively in small, isolated tribes that are governed by the tribes' rigid social hierarchies. Which is to say, even great apes have a civilization, and probably they did long ago.

Seen through this lens, civilization is not, as Freud thought, a recent invention that prevents humans from acting out their most base impulses. Rather it's an expression and refinement of our early animal culture that, even today, helps us be our best animal selves.

Rather convincingly, Morris argues that the difference between tribal homo sapiens sapiens and today's homo sapiens sapiens is merely that there are now about 200,000 - 300,000 times more people per square inch than there were way back when. (Actually, he said "100,000 times more," but Earth has become far more populated since 1969.) According to Morris, crowding, not human nature, is the primary problem disrupting modern behavior. Morris likens the crazy behaviors of some of today's humans to the disturbing, introverted, anti-social behaviors of caged wild animals. Caged chimps smear their excrement, pluck their hair, and bite themselves, sometimes fatally. So can especially harried humans, sometimes.

If Morris was right instead of Freud, the idea that ethics and morality are somehow a result of civilization is hogwash. Instead, they reflect our bestial past. In which case people who extol decency, righteousness, and honor as uniquely human qualities have got their facts topsy turvey. What's more, if civilization itself is characteristic of primates in general, there is no such thing as "pre-civilization" or "pre-culture" regarding human heritage. Any search to understand how culture has enriched or diminished humans' experience of intimacy shouldn't bother comparing non-human primate behaviors with human ones.

It might be more helpful to look at intimate behaviors across today's living cultures. But "intimate behaviors" is too broad a category to easily consider. Even "kissing" is, because there is a splendid variety of them.

Kissing

A point on which Morris and Freud agree is that kissing may have originated in the mother-infant bond. When adults kiss their infants—or each other—they may be re-exciting the pleasure they once took in being nursed. Morris thinks they may also be re-exciting the pleasure of receiving pre-masticated food spit lovingly into their mouths. Regarding bottle babies, plenty of whom enjoy kissing—well, it may be that, evolved as we all are from the great apes and their eons of nursing love, we are all hard-wired to like it.

Alas, just as the first nursing human baby neglected to take pen in hand and make note of his or her pleasure at taking in nutrition, no one recorded the pleasure of the first romantic kiss. We do know, however, that lips were eroticized no later than about 5,000 years ago. The Sumerians decorated their lips and eyes with crushed gemstones. Roughly 3,000 years later Cleopatra painted her lips with red pigments squeezed from mashed-up bugs.

Red has survived in lip color ever since.

Why Red?

The color itself seems to be of primal importance to humans. In research published in 1969 anthropologist Brent Berlin and linguist Paul Kay analyzed 110 languages and found that all had names for black and white. Whenever one of the studied cultures had a name for a third color, the color named was red.

Research psychologist Andrew Eliot and social psychologist Henk Aarts later reported that, in their lab, the sight of the color red enhanced the force and velocity of adults' motor output. They speculated that, to prehistoric humans, that color signaled danger. But women hardly paint their lips red in order to instill fear in men. Might red also signal berries? Fresh meat? And, if so, does red confer a mating advantage? Does it trigger hunger, for example, in both the gut and the groin?

Desmond Morris thinks it's simpler than all of that. He suggests that the response of male humans to the color red comes from humans' great ape days, when females had external, swollen, red labia during estrus. As pre-humans evolved and stood upright and the labia became harder for the males to see, more than just breasts and hips on women changed. Women's lips became extroverted, turning out instead of in. Indeed, so did the lips of all humans. Today, the lips of boys become thinner when they reach adulthood. But girls' lips don't. Theirs stay plump until menopause.

Morris is so enamored of his idea of the origin of red, extroverted lips that he calls them "super labia."

It's a great term. Whatever you think of Desmond Morris as a theorist, you have to admit he's a whiz at catchy names.

An Intimate Culture Clash

Perhaps the first literary reference to human kissing is in Vedic Sanskrit texts from 1500 BC.

The Atharva Veda includes a reference to smelling with the mouth. The Kama Sutra devotes a whole chapter to kissing.

Now, heaven forbid that a young girl from ancient Hindu culture attempts to interpret the kiss of a young 21st century man from North America. Just look:

Here's how North Americans kiss now. This list of 7 tips is from Bustle.com.

—Try an "out of the blue" kiss.
—Use a lot of tongue.
—Make them want more.
—Focus on the lips.
—But don't stick just to the lips.
—Talk about it.
—Remember—Location, location, location.

Here are a few of the many descriptions of kisses from the *Kama Sutra.*

—Throbbing kiss: "When our lower lip moves our lover's lower lip but not the upper."
—Pressed kiss: "When we press our lover's lower lip with much force."
—Touching kiss: "When our tongue touches our lover's

lip and, closing our eyes, our hands touch our beloved's."

What's perhaps more striking than the difference in kissing across cultures is the literary shift over the centuries from genteel descriptions of a gracious mating dance to somewhat pushy guidelines.

And while it may be easy to imagine that cultural norms about kissing would shift profoundly over 3,500 years, even 120 years can make a difference. The first on-screen, passionate kiss was in a 30-second film made in 1896. It showed a couple kissing and talking intermittently and then fully kissing. Remember, the entire clip of kissing and talking lasted 30 seconds. Still, one observer wrote, "The spectacle of their prolonged pasturing on each other's lips was hard to bear.... Such things call for police interference."

Kissing, Kissing Everywhere

Kissing is rare in the animal kingdom. Yes, giraffes entwine necks. Dogs and cats lick and groom. Some birds nuzzle beaks. So do dolphins. But are those necessarily kisses? And if they are, are they romantic?

Interestingly, common chimpanzees kiss, though not romantically. Most often it's done as a sort of apology or to share tastes (and information about food). But members of an endangered subspecies of chimp—pygmy chimps, which are also called "bonobos"—kiss more frequently than common chimps do. What's more, they use tongues. They are also more likely than common chimps to engage in oral sex. Pygmy chimps have redder lips than common chimps, and the males have larger penises. The males and females mate face-to-face sometimes. And, yes. Pygmy chimps spend more time walking upright than common chimps do.

And humans? If you watch movies, it seems that we kiss almost as frequently as we blow up cars, which is to say "all the time." But romantic kissing in humans is less universal than one might think.

According to a 2015 study of data from 168 largely pre-industrial human cultures from around the world, the people of only 46% of civilizations kiss romantically. The ethnographers from the University of Nevada and the Kinsey Institute defined a romantic kiss as "lip-to-lip contact that may or may not be prolonged … not a passing glance but intentional touching of the lips that is more focused and thus potentially more prolonged." By and large, the researchers found romantic kissing more often in complexly stratified cultures. It's also more common in colder cultures than in equatorial, Sub-Saharan, hunter-gatherer ones.

What's the key difference between the warm weather cultures and the cold weather ones? Is it privacy? Cold weather calls for doors and blankets. Warm weather doesn't.

For what it's worth, Darwin believed that all human cultures either kiss romantically or have "kissing-like behaviours." His list of such behaviors included "the rubbing or patting of the arms, breasts, or stomachs" and even an instance of "one man striking his own face with the hands or feet of another." Yes, it is probably difficult for a North American to see the romance in that last example.

But the term "kissing-like behaviours" may also go a long way to explain an intimate custom of people on the Trobriand Islands near New Guinea. In 1929 anthropologist Bronisław Malinowski reported that, during orgasm, lovers bite off one another's eyelashes.

Back to Freud … and On to Love

Just imagine: How might one of Freud's upper middle class Viennese Jews respond to the kissing-like behavior of a Trobriand Islander? What would she do if her eyelashes were bitten mid-orgasm—other than catch her breath and reach for her notebook, write about what just happened, and, of course, bring it up later when in session with Freud?

Now imagine: Margaret Atwood's fictional character, Offred, fantasizing about really good sex. If you've read *The Handmaid's Tale* or seen the hit television series based on it, you'll know that Offred was repeatedly raped by the Commander of the house to which she was assigned, all in an attempt to fertilize her and have her bear a child that would be taken from her and raised by the Commander and his wife. Given that Offred was participating in the sex very much against her will, it felt neither passionate nor intimate. Neither was it affectionate. Yet, in the novel, no matter how often Offred was forced to participate in this procreation ritual, she actively missed the sex she and her own husband used to have. Offred had come of age in a culture that placed a high value on deep kisses and the other accoutrements of romantic love, and her captors were clearly of a different culture.

Just how typical is Offred? In the early 1990s, most researchers thought that passionate love was unique to Euro-American cultures. But Bill Jankowiak, who decades later would lead the 2015 team examining the universality of romantic kisses, thought to read the folklore of—and ethnographic studies about—166 primarily pre-industrial cultures from around the world. In addition to correlating available data, he and fellow researcher Ted Fisher called the authors of individual ethnographic studies and asked whether they'd seen romantic behaviors in the field.

Jankowiak and Fisher used what they called a standard definition of romantic love: "Any intense attraction that involves the idealization of the other, within an erotic context, with the expectation of enduring for some time into the future." In a paper published in 1992 Jankowiak and Fisher reported that couples in 146 of the 166 cultures surveyed behave romantically. Since then, the researchers have re-evaluated their data and added another 5 cultures to the "romance" count. Which suggests that people in 91% of cultures surveyed had the expectations of romantic love that Margaret Atwood's character Offred did.

But is that necessarily a good thing? The ancient Greeks and Romans both considered romantic love destabilizing. Were they right? Might placing all of one's emotional eggs not in one basket but in a web of intimate community and family attachments lead to a more stable and rewarding life? Might it offer humans more opportunities to rise to the biological imperative to procreate?

In America today, we extol the idea of matches that are made in heaven and that will last forever. But getting romantically jilted by someone you've idealized hurts like hell. Maybe the pain would be less and the suicide statistics lower if young people didn't enter exclusive romantic relationships but instead enjoyed a variety of intimate relationships, some sexual and some not. Maybe then, instead of emotionally crumbling when "ghosted" or "dropped," they would take pleasure from the other many people who love them reliably and well.

Cross-Cultural and Cross-Species Intimacy

Would you believe me if I told you that you are sharing intimate space right now with at least hundreds of microscopic mites? Researchers ar North Carolina State University studying a mite called Demodex describe them as fairly good at getting along with humans. They are on your face, but some are probably in your earwax, too. They come in two varieties. The ones called Demodex Folliculorum are tall and thin. The Demodex Brevis are short and fat. Mites are arachnids, and therefore are close cousins to spiders. Don't worry: Unlike spiders, Demodex won't defecate on you; they can't because they don't have anuses. That said, their abdomens get grotesquely big as their short lives progress. And when they die, they and their excrement degrade right where they land.

But don't even try to get rid of them. If you're over 18 they'll be back in a few days. If you're way older than 18 you have more now than you did in your younger days. We can't feel them, see them, or smell them. They don't bother us in the least, except that sometimes they cause rosacea. And sometimes they make eyelashes itch. They seem to have lived on human faces for at least 200,000 years, which suggests that we don't bother them, either.

But here's the kicker about intimacy: Without complaint from you, their human hosts, every night by the hundreds, thousands, maybe even millions, Demodex mites crawl out of their hiding places. And what do they do? They have sex all over your face. You are a completely passive participant in their obscene acts.

Wait for it. All of that cross-species lovey-dovey stuff is going to happen again. Tonight.

For More Information

Darwin, C. (1872). *The Expression of Emotion in Man and Animals.* London: John Murray. Full text available online at http://darwin-online.org.uk/content/

Despommier, D., Gwadz R. W., Hotez P. J. and Knirsch C. A. (2006). *Parasitic Diseases*. 5th ed. Apple Trees Production, LLC.

Elliot, A. J., & Aarts, H. (2011). Perception of the color red enhances the force and velocity of motor output. *Emotion*, 11(2), 445-449. doi:10.1037/a0022599

Freud, S. (1930). Civilization and its discontents. In Strachey, J. (2001) *The Standard Edition of the Complete*

Psychological Works of Sigmund Freud, Volume XXI (1927-1931), Vintage.

Gilmore, D. D. (1991). *Manhood in the Making: Cultural Concepts of Masculinity.* Yale University Press.

Gregor, T. (1977). *The Mehinaku: The Dream of Daily Life in a Brazilian Indian Village*. University of Chicago Press.
Hogenboom, M. (2015). Why do humans kiss each other when most animals don't? BBC Earth. Retrieved at http://www.bbc.com/earth/story/20150714-why-do-we-kiss

Hsu, C. K. Hsu, M. M. & Lee, J. Y. (2009). Demodicosis: a clinicopathological study. *Journal of the American Academy of Dermatology*. 60(3): 453-62

Jankowiak, W. R & Fischer, E. F. (1992). A Cross-Cultural Perspective on Romantic Love. *Ethnology*, 31(2), 149. doi:10.2307/3773618

Jankowiak, W. R.. Volsche, S. L. & Garcia, J. R. (2015). Is the romantic-sexual kiss a near human universal? *American Anthroplogist*. DOI: 10.1111/aman.12286.

Kirschenbaum, S. (2011). *The Science of Kissing: What Our Lips Are Telling Us*. Grand Central Publishing.

Kligman, A. M. & Christensen, M. S. (2011). Demodex folliculorum: requirements for understanding its role in human skin disease. *Journal of Investigative Dermatology*. 131, 8–10

Koren, M. (2013). Heavy metals, insects and other weird things found in lipstick through time. *Smithsonian Magazine*. Retrieved at http://www.smithsonianmag.com/science-nature/heavy-metals-insects-and-other-weird-things-found-in-lipstick-through-time-50868685/ - yJFmDz7qxJwA8jdg.99

Lacey, N. Delane, S. Kavanagh, K. & Powell F. C. (2007). Mite-related bacterial antigens stimulate inflammatory cells in rosacea. *British Journal of Dermatology*. 157(3): 474-81

Lacey, N. Kavanagh, K. & Tseng S. C. (2009). Under the lash: Demodex mites in human diseases. *Biochemistry* (London). 31(4): 2-6

Malinowski, B., (1929). *Sexual life of Savages In Northwestern Melanesia*: Vol. 1 and 2. Horace Liveright. Retrieved at http://ehrafworldcultures.yale.edu/document?id=ol06-005

Morris, D. (1969). *The Naked Ape*. Bantam.

Morris, D. (2004). *The Naked Woman: A Study of the Female Body*. Thomas Dunne Books/St. Martin's Press.

[No author or date given] Romantic or disgusting? Passionate kissing is not a human universal. *Human Relations Area Files*. Retrieved at http://hraf.yale.edu/romantic-or-disgusting-passionate-kissing-is-not-a-human-universal/

[No author], *The Chap-Book*, Volumes 5-6 (1896). Stone and Kimball. Retrieved at https://books.google.com/

The Naked Truth about the Wives of Tall Men

Are the wives of tall men happier? There's actually a bit of research about that.

Reporting in the March 2016 issue of *Personality and Individual Differences,* South Korean researcher Kitae Sohn of Konkuk University says that survey data representing 7,850 couples in Indonesia show resoundingly that wives of tall men are, indeed, happier. When men are about four inches taller than wives, wives are about 4% happier-and

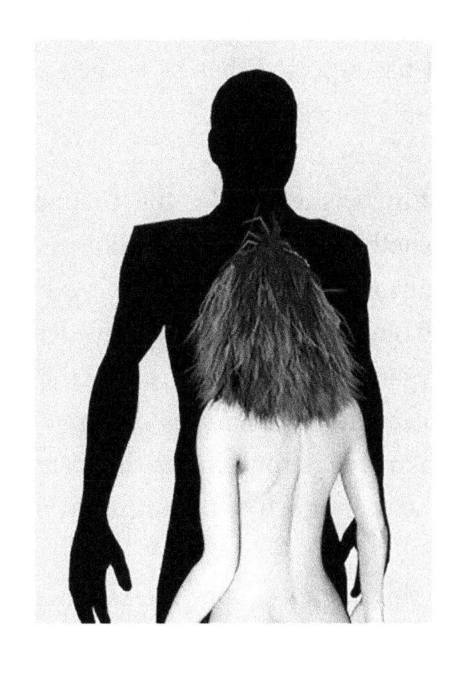

the taller the husbands are, the happier the wives are. It's true, though, that in Sohn's analysis of the data the correlation seems to weaken over the course of about 18 years of marriage until it disappears entirely.

Sohn's findings jive well with the work of other researchers. For example, a 2010 analysis of data collected between 1993 and 1999 showed that 41% of women prefer tall men. In Poland between 1994 and 1996, personal ads representing tall men received more responses. And, while an American commercial dating service called HurryDate doesn't conduct scientific surveys or controlled studies, it has reported that its women clients choose taller men more frequently.

Why women make the mating choices they do, and whether those mating choices make them happy, are burning questions to some evolutionary psychologists. Evolutionary biologists consider the same matters with regards to females of animal species. Commonly, they speculate that females of some species have evolved to favor big, strong males that can protect and provide for them. In much of the animal kingdom, size matters.

Sohn supposes that the same is true for humans—and that, for women, "tall" fits their instinctual preference. Sohn also suggests there may be a second reason that women covet tall men. Giving a nod to the "sexy son" hypothesis originated in 1930 by evolutionary biologist Ronald Fischer, he says that mating with a tall man might help a woman beget tall offspring. By virtue of being tall, any sons would have more opportunities to procreate—and to proliferate his mother's genome.

Sohn acknowledges that his observations about happiness among Indonesians may not apply world-wide. After all, most Indonesians are very short. (The height of the average Indonesian male has been pegged at 5'2". The average female seems to be about 4'10" tall.) Sohn even goes so far as to suggest that Indonesians might be a little overly concerned with height as a result of their own short stature.

With that in mind he suggests that a survey of wifely happiness in Northern Europe might be helpful as a contrast to his work with Indonesian data. In which case I nominate the Dutch as a study population.

According to an April 5, 2004 *New Yorker* article by Burkhard Bulger, the Dutch haven't always been the soaring behemoths we now know them to be. Burger tells a story about visiting a village where Vincent Van Gogh once lived. Apparently, Van Gogh was diminutive, as was typical for Dutch men of the late 19th century.

"I was shown the tiny alcove where the painter probably slept. 'It looks like it would fit only a child,' ... the [building's] current owner, told me."

Bulger reports that, in Van Gogh's day, the Dutch were among the smallest people in the known world. These days, as a result of profound dietary changes, they number among the world's tallest.

Just to repeat: The Dutch went from one of the world's shortest people to one of the world's tallest in about 120 years. That's only six generations—no doubt too short a time for evolutionary prejudices written in DNA to change.

So, I have to ask: How happy are Dutch wives? Anyone have some research money?

Game on.

For More Information

—Sohn, K. (2016). Does a taller husband make a wife happier? *Personality and Individual Differences* (91) 14-21.

—Ronald Fisher, *The Genetical Theory of Natural Selection*, The Clarendon Press (UK), 1930.

—April 5, 2004, Burkhard Bulger's "The Height Gap" in *The New Yorker*.

SCIENCE AND LUST

Men in Red

Men, be honest. Do you have the Tinder app on your smartphone? You know, the one that lets you anonymously browse potential matches near you, swiping right if you're interested and left if you're not? Tinder is popular among the casual hookup crowd because it emphasizes privacy—and requires no emotional investment on the part of the swiper. Who your potential partner is and who you are—well, those nuances are not that important. Where the two of you are and what you each look like are. Or, as a reporter for the *London Telegraph* put it, Tinder "has slimmed down the emotional, cognitive and financial investment required by the virtual dating process to one simple question: 'Do I want to do you?'"

If you want to use Tinder to do and be done, you'll probably want to gussy up first in front of a mirror. And so here's a grooming tip for you men.

Wear red.

You know those red power ties you wore a few years ago? They're probably still at the back of your closet. Put one on; take a selfie, and post it. And when you do, think about how male birds display their red feathers to great mating effect. Feeling lucky now?

I say all of this because an international team of biologists and psychologists have reported that women may be sexually drawn to red on a man—or at least to red around a man. Four hundred and one female university students in Slovakia participated in a study in which they were shown black and white photos of men. The photos were framed with either a grey border or a red one. The women were asked to assess the men's sex appeal and general likeability. Men surrounded by a red border greatly outscored ones surrounded by a grey border.

Oh, yes. Except. Ahem.

Well, it's just that the study's preliminary results actually suggest that women's preference for "men in red" depended on where they are in their menstrual cycle.

It seems that they have to be fertile.

"Whoa, Nellie," right? I'm thinking that's what men who use Tinder to find sex without intimacy are shouting right now. "Red attracts fertile women?????" For a man keen on avoiding genuine intimacy, bedding a fertile woman and becoming a father is probably not on this week's bucket list.

Well, such men will probably be happy to know that this study's authors have flagged their results as preliminary, cautioned their readers about drawing conclusions, and called for further research. You see, it seems that there is a huge open question in the scientific community about how to calculate probable fertility. In this study, the researchers used two common methods: counting forward from the onset of last menstruation and counting backward from onset of menstruation. The women who had been assessed as fertile by researchers counting forward were more likely to find red sexy. Women whose fertility had been assessed by backward counting were not more attracted to red.

With that in mind, this study's authors cautioned about any rush to interpretation of their results. But nowhere did the study authors suggest caution in the dispensation of grooming tips. And so here's one for men who don't want to become fathers right away.

Maybe you'd look better in grey.

For More Information

Prokop, P., Pazda, A. D., Elliot, A. J. (2015). Influence of conception risk and sociosexuality on female attraction to male red, *Personality and Individual Differences (87) 166-170.*

The Human Ape

I write on the morning of November 29, 2017, knowing that, by tonight, any list I could make of powerful men who stand accused of sexual harassment or assault might be obsolete. (NOTE: NBC's Matt Lauer was added as I typed this.)

Source: Senator Al Franken's press kit. Franken was first accused on November 16, 2017 of forcibly kissing a woman, KABC anchor Leann Tweeden.

The accusations against some of the accused men don't

surprise me. But those against Al Franken do. The LGBTQ community, Planned Parenthood, conservationists, and The American Civil Liberties Union all have a friend in Franken.

Franken admits to the allegations by Tweeden. He has called his behavior disgusting.

Damn it. Why do even the good guys act like apes?

They actually are apes.

Ask any ethnologist or anthropologist and he or she will assure you that humans are Great Apes. We share about 98.8 percent of our DNA with chimps and bonobos, which together make up the pan-species.

This means that humans have the same reproductive imperatives that animals in general—and pans in particular—have. The job of the male is to spread his seed as widely as he can. The job of the female is to be selective about whose seed she receives.

Which is to say that humans' biological blueprint lays out dramatically different sex roles for men and women. Men may not think in terms of fathering as many babies as possible, but their hormones make them go whole hog for copulation at almost any opportunity. On the other hand, women's hormones lead them to be choosy and even to recoil at times in disgust.

And therein lies one explanation for the many recent headlines about sexual misconduct by men. Powerful men behave like the animals they are. At the same time, the rules of the game in our society have been constructed mostly by men. This means that, when women exercise their animal right to say, "No," men don't necessarily have to listen.

Our animal nature is undeniable. But is beastly behavior inevitable?

The answer may lie in a look at the animal side of the evolutionary tree.

Very roughly speaking, single celled animals begat fish, who begat land animals, who begat primates, who begat **Great Apes** when they diverged from gibbons about 15 million years ago.

Then **orangutans** diverged about 13 million years ago. Orangutan females are a fairly glee-less lot. Because every adult in the species lives alone, females are easily victimized. In the early 1980s, University of Michigan anthropologist John Mitani closely observed three females. He witnessed 179 copulations over a six-month period. A male initiated all but one, and often the first copulation between any given male and female was a rape.

In *Demonic Males*, a 1996 book he co-authored with writer Dale Peterson, biological anthropologist Richard Wrangham noted that, once an orangutan male has raped a particular female, she is, in general, more acquiescent to him during later encounters. Orangutans are smart. Wrangham guessed that females quickly learn that resistance is futile.

Gorillas split off from the orangutan species about 10 million years ago. They live in groups, which may account for why outright rape is rare.

But here's female gorillas' awful truth. They don't just get impregnated by a male and go on to raise their progeny; they mate long-term. And sometimes, when a male gorilla wants to attract a female away from her current mate, he kills her infant. Afterwards, she willingly runs off with him.

What is it about infanticide that makes it a sexual come-on? Wrangham has speculated that, in gorilla society, a female and her young are probably safest when under the protection of the biggest, baddest gorilla around. When a female's infant is murdered, she knows without a doubt that she has found a truly mean male. So she "trades up" by pairing with him. Yes, he killed her baby. Yes, he can do her future children just as much harm—but he will probably protect them instead because they will be his children.

Pans (**chimpanzees** and **bonobos**) developed into their own species about six or seven million years ago. Male chimpanzees brutally dominate females. Rape is common. But fertile females are often willing sexual partners.

Indeed, fertile females' promiscuity is impressive. Sometimes a female will mate with all of the males in the group in a single day. Wrangham has suggested that this may be in order to create confusion about paternity. Chimps, like gorillas, practice infanticide. If virtually any male in the group could be the father of a particular child, perhaps no male in the group will kill that child.

Most male primates know when a female is fertile because her labia become red and swollen. For chimps, it's no different. Estrus lasts only two to three days out of the approximately 30-day menstrual cycle. So when a female shows the fertility sign, males are all over her and, by and large, she is under them.

For bonobos, though, the signs of estrus extend well

beyond when a female is actually fertile. Maybe that's why males don't respond all that eagerly to swollen labia. They really don't need to; they can easily get around to having sex with any fertile female anyway.

Both male and female bonobos have sex freely and joyfully as often as they like and with virtually anyone they like. Homosexual encounters are as frequent as heterosexual ones. A male might rub his scrotum against the buttocks of another male. Two males might engage in what ethnologists have dubbed "penis fencing." The males hang face to face from branches and, while dangling, rub their erect penises together. Bonobos' unusually large clitorises facilitate female-on-female sex. When two females have sex, one stands and lifts the other. Hugging, they rub their clitorises laterally against each other while they grunt and squeal. Then they clasp each other even more tightly and shudder.

In general, though, both male and female bonobos are inventive about sex. Unconfined to any "script" they can engage their mouths, toes, and fingers virtually anywhere they fit and feel good.

Given the frequency and apparent freedom of bonobo sex, would it qualify as "meaningful" in the way that human mothers often warn daughters sex is supposed to be?

Actually, it does. Bonobos use sex to create alliances that can see them through hard times. Powerful, protective female dyads and gangs are formed that way.

The males never really get around to forming pacts, but they do enjoy themselves greatly—if fleetingly. On average, bonobo sexual encounters last only 13 seconds.

Females are dominant in bonobo society. When a male attempts to be sexually coercive, "no" means "no," and a gang of very angry, sexually beholden to no one females stands at the ready to defend his intended victim.

Hominids separated into a new species at about the same time as pans did. Anatomically modern humans—Homo sapiens—came on the scene between 100,000 and 200,000 years ago.

Are modern human males as sexually animalistic as their ancestors? Today's rape statistics vary by culture and by data collection method. But, according to the Centers for Disease Control, nearly one in five American women report being raped at some time in their lives. While infanticide happens all too frequently in America, it probably isn't usually instigated as a mating come-on. Which is to say that American women fare far better sexually than female orangutans and gorillas.

The Human Ape

Humans evolved much more quickly than any of their primate cousins or ancestors. All pans, for example, are pretty much the size and weight their ancestors were when the pan line split off from the Great Ape evolutionary branch. But humans look very different than they did six or seven million years ago. And some can do cool tricks like flying to the moon.

One common explanation for the rapid evolution of humans is that hominids became a separate species because their territory became drought stricken. Those who survived the great drought did so by figuring out to stop lazily counting on fruit to appear. Instead, they dug for roots. Ultimately, as their part of the jungle became woodland, they incorporated a wide array of new foods into their diets. Early humans learned to dig, but they also learned to hunt, hammer with increasingly refined tools, and cook. As they developed the technologies to accomplish necessary tasks, form followed function. Their bodies changed and their brains got bigger.

These days, while orangutans, gorillas, and pans continue to pluck jungle fruit, we humans invent and re-invent our technologies and even our cultures almost *ad nauseum*. We devise laws and we codify punishments for a spectacular array of specific misbehaviors.

"I went at her like a bitch."

Our president said that. That may be because he's a particularly ape-ish Great Ape, as Harvey Weinstein probably is. Oh. And Al Franken.

But to what extent should we accept the idea that all apish sexual behavior is equally loathsome? Is Franken a Weinstein? The most pressing problem we want to address isn't gross insensitivity, is it? It's raw abuse of physical, economic, and emotional power to sexual ends.

So, what about clumsy and sometimes disgusting approaches perpetrated by men who, like bonobos, seem to assume that everyone is up for sex 100 percent of the time, and stick their tongues and hands in all the wrong places? What about Franken's admitted and decidedly bonobo-like attempts at 13 seconds of erotic flame?

Rose is a Rose

A 2014 paper from the University of North Dakota reported that 31.7 percent of college men surveyed in 1980 allowed that, in a consequence-free situation, they'd force a woman to have sexual intercourse.

At the same time, only 13.6 percent said they would rape a woman.

Look at that 20-percentage point spread. What's the difference between forced sex and rape?

Of course, there isn't one. Helpfully, the researchers noted that, "When, instead of using labels like 'rape,' survey items ask questions about behaviors, more men admit to the behavior that the label describes. For example, "Have you ever coerced somebody to intercourse by holding them down?' elicits more 'yes' answers than 'Have you ever raped somebody?'"

This reminds me of the excuses we've heard recently from the lawyers of accused men. Often they claim that their clients are aghast to learn that not all of the sex they've enjoyed was consensual.

Even Al Franken initially said, "I certainly don't remember the rehearsal for the skit in the same way, but I send my sincerest apologies to Leeann."

In his later, full statement, Franken got more to the point.

"Over the last few months, all of us—including and especially men who respect women—have been forced to take a good, hard look at our own actions and think (perhaps, shamefully, for the first time) about how those actions have affected women.

"For instance, that picture. I don't know what was in my head when I took that picture, and it doesn't matter. There's no excuse. I look at it now and I feel disgusted with myself. It isn't funny. It's completely inappropriate. It's obvious how Leeann would feel violated by that picture. And, what's more, I can see how millions of other women would feel violated by it—women who have had similar experiences in their own lives, women who fear having those experiences, women who look up to me, women who have counted on me."

The eloquence of that apology helps a bit with my feelings of profound disappointment about Franken. And thinking of him as a boorish, occasionally unzipped, but not sexually coercive Great Ape makes me imagine that he deserves strong censure, but not the loss of his Senate seat. Perhaps there's a significant difference between a dunderhead's shenanigans and those of men who are violent or use their positions of power to get what a woman would rather not give.

I do wish, though, that we could get a horde of our bonobo sisters together to teach Franken a lesson. With luck, what the media do to him as he returns to the Senate will substitute nicely. And, of course, the Senate's Ethics Committee will dig, and that's all good.

We are left for the time being with Franken's full-throated apology. If nothing else it has given American men an opportunity to examine their apish behavior and find words with which to reach out to women they've harmed.

I only hope that, in doing so, men everywhere take no cues whatsoever from their bonobo cousins.

How do bonobos apologize, you might ask? They kiss and quickly initiate sex.

For more information:

Dale Peterson and Richard Wrangham, *Demonic Males: Apes and the Origins of Human Violence* (Mariner Books, 1996).

The Erotic Analysis of Anna Freud by Her Father, Sigmund

Once Upon a Time (Really)....

Anna Freud was inarguably her father Sigmund Freud's favorite child. Sigmund's intellectual descendants readily concede that point.

What many Freudians don't concede is that there was anything inappropriate in Anna's relationship with her father. For the longest time, most didn't have cause to.

Anna Freud was born in 1895 and lived until 1982.

Unlike her two sisters and three brothers, she never married. Instead, beginning in her mid teens, she became Freud's pupil of sorts. By her early adulthood she was his nearly constant companion and, eventually, his closest collaborator. As Freud's cigar habit contributed to cancer, and as cancer surgeries increasingly debilitated him, Anna began tending physically to her father day and night. This arrangement continued until Freud's death, when Anna informally assumed the mantle of the head of the psychoanalytic movement and bore his psychoanalytic seed into the world. Ultimately, Anna became a pioneer in the field of child psychoanalysis, adding immeasurably to Western understanding of childhood development.

It wasn't until the late 1960s that Paul Roazen, a political scientist and historian of the development of psychoanalysis, stumbled upon an immense skeleton in the Freud family closet. Even though Freud defined analysis as an erotic relationship laden with transference and countertransference, Freud analyzed Anna.[1] [2] He did so for two periods, one beginning in 1918 and one beginning in 1924.[3] Roazen published his discovery in 1969 in *Brother Animal* (Knopf), a book that raised serious questions about Freud's role in the bizarre suicide of one of his most brilliant pupils.[4] The book was not heartily embraced by the psychoanalytic community. In fact, in its November 5, 2012 obituary for Roazen, *The Washington*

Post quoted Anna Freud as saying, "Everything Paul Roazen writes is a menace."

So while, beginning in about 1969, some people within the psychoanalytic community surely knew about the illicit analysis, for the most part, that community treated Roazen's views—and news—with scorn. It wasn't until 1988 that news of the improper analysis officially "broke" within the psychoanalytic community itself. *Anna Freud: A Biography* (Norton) is the only authorized biography of Anna. In it, psychoanalyst Elisabeth Young-Bruehl described the analysis without commenting on its propriety or lack thereof. On the other hand, that same year in *Freud: A Life for Our Times*, Peter Gay (a Yale-affiliated historian with no professional allegiance to Freud's ideas) called that particular psychoanalysis "a most irregular preceding," and Freud's decision to analyze Anna, "a calculated flouting of the rules he had lain down with such force and precision."[5]

Indeed, again and again Freud had cautioned his colleagues about the rules. "Never, ever try this at home" is essentially what he said about psychoanalysis and families.

Questions for Freud

Both Gay and Young-Bruehl named the analyses' primary

topic: Anna's masturbation fantasies, which were frequent, violent, and masochistic. Anna had been masturbating to what she called "beating fantasies" since about age six. The earlier fantasies took different forms and had vaguely defined protagonists and antagonists. Once she was an adult, the fantasies were clearly about Anna. She imagined herself a young man being held captive by a knight who tried to force her to betray "family secrets." Even though the youth never made a whole-hearted attempt to escape from the knight, he refused to blab. The youth was always beaten by the thoroughly enraged knight.[6]

When Freud first analyzed Anna, he seems to have done so with the objective of relieving her of her habit of masturbating—a habit that he considered masculine in nature and therefore dangerously inappropriate for females.[7] [8] [9] But was Freud concerned about more than just the masturbating? Anna had not married; she'd never even dated. In the beating fantasies that she discussed with her father she played the role of a male (albeit a male in a homoerotic relationship with another male). Was Freud also concerned about a general tendency in Anna towards masculinity? And was Anna indeed struggling with questions of sexual preference when she first entered analysis with her father?

In 1922 Anna terminated her analysis; the record is unclear

about why she chose to do so. Perhaps the frequency of Anna's masturbating had diminished. Regardless, by 1924 she was again masturbating regularly—and enjoying it immensely. "I am impressed by how unchangeable and forceful and alluring such a daydream [of the young man held captive by the knight] is, even when it has been—like my poor one—pulled apart, analyzed, published, and in every way mishandled and mistreated," Anna wrote to her friend, the novelist and femme fatale, Lou Andreas-Salomé. "I know that it is really shameful ... but it [is] very beautiful."

In 1924, Anna reentered analysis with her father.

Also in 1924, Anna met an American woman, Dorothy Burlingham, heir to the Tiffany fortune. Almost immediately, Dorothy and her four children took up permanent residence in Vienna. Soon enough they moved into an apartment in the same building as that of the Freud family. Anna moved many of her personal items out of her family's apartment and into Dorothy's. Dorothy and Anna began vacationing together and bought a small house in the country.[10] Eventually Dorothy began referring to Anna as the second mother to her children.[11]

Almost any concerned father of Freud's day would have hoped that his daughter would marry and have a family.

So Freud may be forgiven for at least wanting Anna to be analyzed in 1924, given the fact that she was about to turn 30, her biological clock was ticking, she was masturbating again and aplenty, and her sexual fantasies were not classically "boy meets girl, boy marries girl, girl has children" ones. Freud may also have been especially interested in analyzing Anna once Dorothy entered the picture and the apartment building.

But given that Freud knew that analysis was always erotically charged, why didn't he refer Anna to a colleague for analysis? By his own theorizing, if his daughter were a lesbian, mistakes that he had made as a father were the cause of that trouble.[12] [13] Was Freud too worried about his personal reputation to let a colleague talk frankly with Anna? Was he hoping that, as Anna's analyst, he could quietly rectify any "problems" he had "caused"—and help her refuse a life that would speak embarrassingly about his failings?

And just how erotic did things get in the analytic sessions between Sigmund and Anna Freud?

No doubt, an answer to the question about incestuous overtones in the father-daughter psychoanalytic relationship would be easier to improvise if Anna's sexual fantasies had demonstrably changed over the course of the analyses in a way that invited speculation. But they didn't. Anna continued on with her "young man meets knight, young man gets imprisoned by knight, young man doesn't actually try to escape from knight, young man gets beaten by knight and Anna has an orgasm" fantasy life.

Anna, however, was hardly the only significant party to the analysis. Surely her father's thoughts, feelings, and fantasies were tugged hither and yon in all of that transference and countertransference. Does evidence exist suggesting how Freud himself was affected? Changed?

Certainly we know nothing about his sexual fantasies during that period or any changes in them. We do know, however, about his theories about women, and those did change, significantly, during the years of his analyses of Anna. Or at least two theories did.

During the six years in which Freud analyzed Anna, he redefined penis envy as the major factor in a woman's sexual development and he redefined masochism as an expression of female nature.

Penis Envy

Way back when Anna was 10 years old, Freud had first theorized about penis envy, but back then he had framed the concept rather innocently. In 1905's *Three Essays on the Theory of Sexuality* Freud had sounded neither pejorative nor terribly informative about a girl's supposed desire for a penis of her own. He said that all little boys assume that everyone has a penis. When presented with evidence to the contrary they deny the absence that they behold. Little girls, on the other hand, do not resort to denial. They immediately recognize that a boy's genital is bigger than a girl's and they want one more like the one that boys have. Penis envy as described in 1905 was much like the envy that any child with a small scooter might have for another child with a large tricycle. For a child size always matters.

Even as a little girl, Anna had been a precocious thinker, so independent that her father's letters to friends were proudly speckled with anecdotes about her feral ways. As a young woman, she remained unconventional. Other girls looked forward to lives as homemakers. Anna did not. She wanted to know about analysis. She wanted to meet analysts and breathe analysis. She wanted to discuss ideas.

However, once Freud took 23-year-old Anna into analysis and the inevitable web of sexual attractions were woven, Anna grew emotionally dependent on her father, more than she had ever been before. While they were briefly separated in 1920, she wrote to Freud, "You surely can't imagine how much I continually think of you." Around the same period, Freud's letters to friends began including concerns about Anna's increasingly unshakable attachment to him. In 1921, he wrote to his Berlin colleague, Max Eitington, "I wish that she would soon find reason to exchange her attachment to her old father for a more durable one."

It was in 1925 that Freud published an elaboration of his original theory of penis envy. He said, essentially, that the moment at which a girl first discovers her lack of a penis is a moment of ineradicable psychic trauma. From that single moment on the girl will want a penis. As she matures, however, she will realize that she can never under any circumstances grow one. Hoping, then, for second best, she will begin to desire her father's penis. Because she knows that incest is taboo, the girl's desire for her father's penis will be wrapped in shame. She will eventually sublimate her desire for her father's penis into a desire for a child. To fulfill her obsession to have children throughout her childbearing years she will need to secure and retain a man.[14]

Freud believed that boys build their moral sense from a fear that their fathers might castrate rather than spank them. A girl, however, has no penis to lose to her father's rage and therefore no good incentive to develop moral virtue. Whatever virtuous behavior she will manage to exhibit will derive from her quest to catch and maintain the man who can provide her with cute and cuddly penis substitutes. Or so said Freud, roughly.[15]

By Freud's own understanding of the erotic tangle that psychoanalysis creates, each session of Anna's six-night-a-week psychoanalysis was one in which she and her father/analyst sexually desired each other. There seems no reason not to assume that each of them conducted themselves admirably in spite of the abundant opportunity the privacy of analysis gave them to transgress almost any boundary imaginable. Rather, there is every reason to assume that they went through whatever machinations were necessary to avoid acting on whatever desire they felt. It seems probable that their actual behavior was impeccable, and this in spite of the fact that the nightly topic of conversation (the youth, the knight, the youth's curious failure to escape, and, oh, that inevitable beating) probably fed the general level of agitation in the room.

Plain English: However innocent its beginnings, Freud's theory about penis envy and a girl's overwhelming desire for her father's penis may be based only minimally on observable phenomena in girls and young women in general. More profoundly, it might be about Freud's own daughter and Freud's own penis.

Masochism

In 1905's *Three Essays on the Theory of Sexuality* Freud also first discussed masochism, observing that certain people require physical or mental pain in order to achieve sexual satisfactions. Unequivocally he called masochism a perversion.

Then, in a 1919 paper called "A Child is Being Beaten," Freud normalized masochism, suggesting that masochistic elements in childhood sexual fantasies can be sexual representations of underlying feelings of guilt. Freud based "A Child is Being Beaten" on his analysis of the masochistic fantasies of two boys and four girls. However, one child's case material made up the lion's share of the paper's documentary evidence.

That child was Anna. The fantasies were the childhood versions of her adult beating fantasies. We know this because, three years later, Anna described the same child and the same fantasies in "Beating Fantasies and Daydreams," her first psychoanalytic paper ever. (In "Beating Fantasies and Daydreams," Anna referred to the child as a patient of hers. However, as psychoanalyst and historian Elisabeth Young-Bruehl pointed out in her authorized biography of Anna, the conceit is transparent. The "patient" must have been Anna herself, for it would be another six months before Anna began psychoanalyzing anyone.)[16]

Then, in 1924, Freud elaborated on masochism, suggesting for the first time that it is quintessentially feminine to find pleasure in pain—indeed that masochism is "an expression of the feminine nature."[17]

To his credit, there is no evidence that Freud based his 1924 idea of feminine masochism mostly on his analysis of Anna. Regardless, he did write it during his analysis of her. So it seems fair to ask: To what degree did his analysis of and hopes for Anna convince Freud that masochism is characteristically female? Keep in mind that Freud, a self-described "conquistador" of the inner world,[18] required Anna to recline on the couch six nights a week while he psychologically teased her. Keep in mind also that she complied.

Keep in mind that the young man in Anna's beating fantasies never once tried to escape the pleasure of the knight.

All the King's Horses and All the King's Men

Shortly after Anna's second bout of analysis came to an end, she went on to cohabit happily ever after with Dorothy.

Evidently, even the king of psychoanalytic persuasion could not dictate to his daughter whom and how she would love. Freud seems to have grudgingly accepted Anna and Dorothy's relationship. In his correspondence with friends he referred to them both fondly as "virgins" but, as the years passed, stopped wishing for the day that Anna would marry. Following Freud's suit, over the years, friends and family and even Freud's most doctrinaire adherents acknowledged Anna and Dorothy's relationship as intimate and exclusive. However, no one but the maid ever hinted that it was sexual. (According to Jeffrey Moussaieff Masson, former Projects Director at the Freud Archives, Anna and Dorothy's maid told him that they "shared a bedroom" from time to time.[19]) And in an interview with Freudian psychoanalyst Isaac Tylim, Dorothy's grandson Michael reported that his father (Dorothy's son Robert) left a deathbed note spilling the family beans by acknowledging that Anna and Dorothy were, indeed, lovers.[20]

Notes

[1] "Perhaps the most extraordinary illustration of Freud allowing himself privileges he might have condemned in any other analyst was his analyzing his youngest child, Anna. Freud analyzed Anna in the period at the end of World War I. In letters Freud was quite open about this

analysis, and it became a pubic secret to a small group of his inner circle. From Freud's point of view there were probably some good reasons for doing what he did. But considering all the discussion in later years about what constitutes proper psychoanalytic technique, Freud's liberty in analyzing his own child makes one skeptical of ritualism in therapy or training." Paul Roazen in his 1969 book *Brother Anima*l (Knopf), page 100.

[2] From page 433 of Elisabeth Young Bruehl's 1988 *Anna Freud: A Biography* (Summit): "Roazen interviewed some associates of the Freud family for his 1975 book, *Freud and His Followers*, and Anna Freud had tense exchanges with her friends as she tried to find out who had discussed with Roazen such matters as her own analysis with her father, which she had refused to discuss when asked about it by historians."

[3] In his essay, "'A Child is Being Beaten: A Clinical, Historical, and Textual Study," in the 1997 book *On Freud's "A Child is Being Beaten"* (Yale University Press) Patrick Joseph Mahoney says, "Without doubt, Freud was stimulated to explore beating phantasies because they were central in the dynamics of his daughter, whom he began analyzing in October 1918. (He finished her first analysis in the spring of 1922.)" This is on page 49. Elisabeth Young-Bruehl, in her 1988 *Anna Freud: A Biography*,

says on page 81 that the first analysis began in October of 1918. On page 107 she says that the 1918 analysis lasted "nearly four years." On page 122 Young-Bruehl writes that Anna began her second analysis "after a two-year pause," and on page 124 she describes it as having taken place in 1924 and 1925.

[4] *The New York Times*' obituary of Paul Roazen reads, in part: "A more recent book, *How Freud Worked: First-Hand Accounts of Patients* (Jason Aronson, 1995), continued Dr. Roazen's fascination with Freud's breaches of his stated methods and practices. It revealed that Freud had analyzed his daughter, Anna, as well as a friend of Anna's, Eva Rosenfeld, while Eva lived in Freud's household, despite his emphasis on maintaining objective distance between analyst and patient."

[5] *Freud: A Life for Our Time*, page 440.

[6] Rivka R. Eifferman gives a nice summary of Anna's masturbation fantasies in her essay, "The Learning and Teaching of Freud," published as part of Ethel Person's *On Freud's "A Child is Being Beaten"* (1997,) Yale University Press, p. 171.

[7] In his early writings Freud did not declare an attitude toward masturbation. However, he apparently considered

it a treatable problem, for in 1895 he referred a patient, Emma Eckstein, who was a chronic masturbator, to his friend Wilhelm Fliess for nasal surgery. Fliess believed that certain sexual problems could be relieved by surgeries involving the patient's nose. See J. M. Masson's 1984 *The Assault on Truth: Freud's Suppression of the Seduction Theory* (Farrar, Straus and Giroux). See also Freud's essay "On the Grounds for Detaching a Particular Syndrome from Neurasthenia under the Description 'Anxiety Neurosis.'" In it Freud makes reference to Fliess's idea of "nasal reflex neurosis." (I believe this is in Vol. III of Strachey's *The Collected Papers of Sigmund Freud.* See page 90.) Furthermore, in "The Neuro-Psychoses of Defense" (1894) Freud writes about "a girl who suffered from obsessional self-reproaches. Stimulated by a chance voluptuous sensation, she had allowed herself to be led astray by a woman friend into masturbating, and had practiced it for years, fully conscious of her wrong-doing…." (Strachey, *Standard Edition Volume III*, page 55.) Freud does not explicitly condemn masturbation but his attitude toward it is apparent.

[8] In 1932 Freud formalized his attitude toward masturbation. "An overwhelming aetiological importance is attributed by neurotics to their masturbatory practices. They make them responsible for all their troubles, and we have the greatest difficulty in getting them to believe

that they are wrong. But as a matter of fact we ought to admit that they are in the right, for masturbation is the executive agent of infantile sexuality, from the faulty development of which they are suffering. The difference is that what the neurotics are blaming is the masturbation of the pubertal stage: the infantile masturbation, which is the one that really matters, has been for the most part forgotten by them." This is from *Freud's New Introductory Lectures on Psychoanalysis*, 1932, Chapter 5 (edited by James Strachey) and quoted in the 1965 *Freud: Dictionary of Psychoanalysis*, edited by Nandor Fodor and Frank Gaynor (Fawcett Publications). See page 94.

[9] By 1925 Freud's attitude toward masturbation in women had become quite clear. In "Some Psychical Consequences to the Anatomical Differences Between the Sexes" he writes, "But it appeared to me nevertheless as though masturbation were further removed from the nature of women than of men, and the solution of the problem could be assisted by the reflection that masturbation, at all events of the clitoris, is a masculine activity and that the elimination of clitoridal sexuality is a necessary precondition for the development of femininity." Strachey, Volume XIX, 255.

[10] Paul Roazen's *How Freud Worked*, page 97.

[11] The early history of Anna's and Dorothy's almost six-decade relationship is described by Elisabeth Young-Bruehl on pages 132-139 of *Anna Freud: A Biography*.

[12] Diagnosing a female homosexual patient, Freud says "[a]fter her disappointment [with her father], therefore, this girl had entirely repudiated her wish for a child, the love of a man, and womanhood altogether. . . . She changed into a man, and took her mother in place of her father as her love-object" ("The Psychogenesis of a Case of Homosexuality in a Woman," in *Sexuality and the Psychology of Love*, ed. Philip Reiff, trans. Joan Riviere (Collier Books, 1920). See page 144.

[13] Freud writes in "Some Neurotic Mechanisms in Jealousy, Paranoia, and Homosexuality" (1922), "We subsequently discovered, as another powerful motive urging towards homosexual object-choice, regard for the father or fear of him…." Strachey, Volume XVIII, page 231.

[14] Strachey, Voume XIX, pages 252-256.

[15] A few years after "Some Psychical Consequences to the Anatomical Differences Between the Sexes," Freud wrote once again about penis envy. "We ascribe a castration-complex to the female sex as well as to the

male. That complex has not the same content in girls as in boys. The castration-complex in the girl, as well, is started by the sight of the genital organs of the other sex. She immediately notices its difference and—it must be said—its significance. She feels herself at a great disadvantage and often declares that she would 'like to have something like that too,' and falls victim to penis-envy, which leaves ineradicable traces on her development and character-formation and, even in the most favourable instances, is not overcome without a great expenditure of mental energy. That the girl recognizes the fact that she lacks a penis does not mean that she accepts its absence lightly. On the contrary, she clings for along time to the desire to get something like it, and believes in that possibility for an extraordinary number of years.... The discovery of her castration is a turning point in the life of the girl.... The wish with which the girl turns to her father is, no doubt, ultimately the wish for the penis.... The feminine situation is, however, only established when the wish for the penis is replaced by the wish for a child—the child taking the place of the penis, in accordance with the old symbolic equation.... Her happiness is great indeed when this desire for a child one day finds real fulfillment; but especially this is so if the child is a little boy who brings the longed-for penis with him." *Freud: New Introductory Lectures on Psychoanalysis*, 1932, Chapter 5, quoted in *Freud: Dictionary of Psychoanalysis*, edited by Nandor

Fodor and Frank Gaynor. Greenwich, CT: Fawcett Publications, 1965, page 116-117.

[16] From Elisabeth Young-Bruehl's *Anna Freud: A Biography*, page 104: "But it is at least clear from her various correspondences that [Anna Freud's] 'Beating Fantasies and Daydreams' was modeled—in general, if not in complete details—on her own case, and her essay's descriptive framework is identical with the one that applies to two of the female cases in Freud's essay."

[17] "The Economic Problem of Masochism," in Strachey Volume XIX, 159-170. "Expression of the feminine nature" is found on page 161.

[18] In 1900 Freud wrote to his friend, Wilhelm Fliess, "I am actually not a man of science at all. . . . I am nothing but a conquistador by temperament, an adventurer." *Encyclopedia Britannica Online* at http://www.britannica.com/eb/article-22606.

[19] Jeffrey Moussaieff Masson's *Final Analysis*, page 158.

[20] Tylim reported this at the 50th Congress of The International Psychoanalytical Association in a session called "Dining with Anna Freud." In that session he also read from two letters written by Anna to Dorothy and

loaned to him by Dorothy's grandson, Michael. Most of the correspondence between Anna and Dorothy is held by The Sigmund Freud Archives and is not available for perusal by researchers.

Sexual Strategies of the Female Narcissist

In today's rape-aware climate, allegedly predatory males aren't given an easy pass: Just ask Harvey Weinstein, Donald Trump, Bill Cosby, Bill Clinton, and Woody Allen. I was about to celebrate 24 hours of not noticing a headline about any one of those men in a major news source when I saw a new one on the website of *People*... and then another on *Vanity Fair*.

Is the behavior that "Weinstein & Co." stand accused of rare among famous men? And are men, famous or not, unique in employing sexually coercive behavior?

Apparently, it is not—and they are not.

According to research psychologists from the University of Liverpool, women are often sexually coercive. But rather than rape, they use emotionally manipulative tactics like threatening blackmail or vowing to harm themselves. Or they resort to seduction via drugs or alcohol, *à la* the Cosby allegations.

According to the Liverpool team's paper, "The Ultimate Femme Fatale," the women to watch out for are, not surprisingly, narcissistic. (Those are the men to watch out for, too.) The paper cites study after study showing that, in both men and women, narcissists' inflated self-importance, deep need for admiration, and lack of empathy are statistically associated with persistent sexual persuasion, coercion, and aggression.

If unwelcome sex is practiced by narcissists of both sexes, why did this study call attention only to women? According to the authors, virtually all previous scholarly investigations of the relationship between narcissism and sexually predatory behavior had been conducted only with men. Theirs, they believe, is the first to include women. Their study sample was 329 adults, most of them undergraduate students at a university in northwest England. Narcissism was measured using a standard personality inventory that also rendered scores for a few personality measures not strictly part of the narcissism portfolio, such as feelings of entitlement and self-perceptions as a leader or authority figure. Sexually coercive tactics were measured by the Post-Refusal Sexual Persistence Scale, a 19-question survey that ranks coercive tactics by increasing severity.

Some highlights:

—In general, the higher either a man or a woman scored on measures of narcissism, the more likely he or she was to report having used sexually coercive tactics.

—Men scored significantly higher than women on measures of narcissism and sexual coercion. But, like men, narcissistic women were more likely than other women to be sexual bullies.

—When men perpetrated coercive sex, they often resorted to physical force. Women, however, were more covert and cunning, using lies and threats.

—Many men with a high sense of entitlement did follow the pattern of covert coercion. On the other hand, coercive men who saw themselves as leaders or authority figures were likely to use any kind of sexual persuasion on the Post-Refusal Sexual Persistence Scale—except if they also scored high on classic measures of narcissism. Men who scored high on that scale tended to favor emotional manipulation and taking advantage of an intoxicated target.

—Narcissistic men and women both reported that sexual refusals excited them, fueling desire and leading them to escalate tame situations into coercive ones.

One may pore over results like this and gab about where Bill Cosby, Bill Clinton, or any number of accused-but-not-convicted newsmakers may or may not fit in. But the news here is not that celebrities and politicians can be narcissists, or even that sexual predators can be. Long before this study, research had turned up solid evidence of all that.

The news is that narcissistic women can be sexually dangerous.

Their tactics differ from those of men. They may inflict guilt and fear more than bruises or broken bones. But it appears that sexual predation knows no gender boundaries when the potential predator has inflated self-importance, a deep need for admiration, and a lack of empathy. And sexual aggression by narcissists of either gender can be difficult to escape—really difficult.

For More Information

Blinkhorn, V., Lyons, M., and Almond, L. The ultimate femme fatale? Narcissism predicts serious and aggressive sexually coercive behaviour in females. In *Personality and Individual Differences*. Volume 87 (December 2015) pages 219-223.

The Gift that Keeps You Giving

Well, I'll be *doo-doo-doo-down-dooby-doo-down damned.* Millions of years of human evolution and about thirty years of formal inquiry have shown that breaking up is hard to do. A new study may show *why, why that is true.*

Research reported during the years 1980-2008 consistently showed that romantic partners are reluctant to watch the trust, love, hope, and time they've invested in a relationship wash down the drain. Indeed, a person's perceptions of the extent of his or her own commitment can become a trap.

But what about when people focus not on their own but on their romantic partner's commitment? The new research, by psychologists at the University of Toronto and UC Berkeley, examined the effect on a romantic partner of perceptions of his or her loved one's investment.

The researchers conducted three experiments. The first involved 108 dating or married couples. One group of couples was asked to think and write about their romantic partner's commitment to them. One group was asked to think and write about their commitment to their partner. A control group was not asked to think or write about commitment. Everyone was asked after the thinking and writing (or not) to rate their own level of commitment to their current relationship. The researchers found that people who had thought and written about their partner's commitment scored as having a significantly higher level of commitment. And this wasn't because they felt guilty about leaving. Post-exercise assessments showed that both men and women had been imbued with feelings of trust and gratitude.

Study 2 asked individuals to take notes daily for one week about their perceptions of their partner's commitment. Data analysis showed an immediate increase in gratitude and in relationship commitment that was still evident nine months later.

In the third study of the series participants noted over a two-week period the frequency at which their partner exhibited his or her commitment. The more frequently an individual perceived commitment, the greater was his or her own commitment three months later.

"Placing resources into a romantic relationship, such as

one's time, energy, emotions, and material goods, may be an effective way to elicit feelings of gratitude from a romantic partner," the researchers wrote, speculating that the resulting gratitude is the tie that binds.

More informally, about romantic investments co-author Samantha Joel suggests that they are "a double-edged sword. They make everyone more committed, which is only a good thing if the relationship is a good thing. So, don't make major investments (or encourage your partner to make major investments) unless you're sure the relationship is fulfilling and you want it to last long-term."

And don't say that this is the end
Instead of breaking up I wish
That we were making up again.

For More Information

Joel, S., Gordon, A. E., Impett, E. A., MacDonald, G., and Keltner, D. (2013). The things you do for me: perceptions of a romantic partner's investments promote gratitude and commitment. *Personality and Social Psychology Bulletin* 39 (10) 1333 –1345.

Improve Your Sex Appeal

Tips from science:

1—Aki Sinkkonen of the University of Helsinki suggests that, over the ages, a woman's belly button has been a sign of her vigor sexually and reproductively. Reviewing assessments of male preference conducted by other researchers he notes that women's innies are very popular. And, he says, ideally those innies are delicate—not too much hooding, and not too cavernous.[1]

2—When bad odors happen to good people, it's decidedly unsexy. Some people stink no matter how often they bathe. In a study at a smell lab in Philadelphia, scientists ran a simple chemical test on 353 people who had "malodor production" that couldn't be explained by poor hygiene, flatulence, or dental problems. They found that about one third of them tested positive for trimethylaminuria (TMAU), a genetically-transmitted disease that thwarts the body's ability to metabolize trimethylamine (TMA), a chemical compound common in eggs, some legumes, wheat germ, saltwater fish, and organ meats. When TMA is inadequately metabolized, it accumulates in urine, sweat, and saliva. The result is a gently radiating, fishy stench that is far from popular in wine bars. TMAU is easily diagnosed with a chemical test that can be run in a medical lab. And that's good, because a diet hygienically "clean" of the foods carrying high TMA concentrations clears the air. [2]

3—Gentlemen think blondes are young. At Poland's University of Wroclaw 360 men were shown pictures of a woman whose photo had been manipulated to display her with blonde, brown, and dark brown hair at 20, 30, and 40 years old. When blonde she was generally judged to be younger. [3]

4—Gentlemen prefer blonde hitchhikers. Researchers

at France's University of South Brittany had five young women rotate between blonde, brown, or black wigs while hitchhiking. Blonde wigs elicited offers of rides 18 percent of the time. Brown and black wigs "stopped traffic" only 14 percent and 13 percent of the time, respectively.[4]

5—By Internet survey, a professor of consumer behavior and marketing at Cornell University's School of Hotel Administration learned that waitresses who are blonde get better tips—especially if they are also busty, skinny, and young. [5]

6—Men: Prepare to toss your toupees. In studies conducted at the Wharton School, subjects looking at photos judged men with shaved heads to be taller, stronger, and more dominant than those wimps with hair. [6]

7—Men Again: If it's a one-night-stand you're looking for, rouge up your facial scars. Scientists at the University of Liverpool found that women looking for short-term relationships prefer facial scars, perhaps seeing them as signs of bravery. They don't want to marry those men, but they do want to have a good time with them.[7]

8—Whatever happened to *Our Bodies, Our Selves*? Surveying 2,451 women, researchers at Indiana University learned that pubic waxing is associated with more positive genital self-image and sexual function. Really?[8]

9—And really again? Psychologists at Lafayette College in Pennsylvania found that college students make no negative assumptions about a woman when told that unshaved body hair is a result of a medical condition. But when they're told that she's hairy because she's a feminist, many rate her as unfriendly, immoral, aggressive, unsociable, and dominant.[9]

10—Publishing about the years 1842-1971 in the *Journal of Nonverbal Behavior*, evolutionary biologist Nigel Barber noted that, whenever the marriage market in Britain got tight for men, facial hair came into fashion.[10]

11—Archeologists in London have found a jar of cosmetic cream from about 2,000 years ago, when Rome occupied Great Britain. It was made of animal fat, starch, and tin. (This represents an improvement over other cosmetics of the time, which were lead-based.) A modern batch made of the same ingredients left faces looking spectacularly white.[11]

12—With the idea that higher status makes them more sexually attractive to richer men, many women spend time and money on clothes and cosmetics. But they would definitely not go to those all that trouble if they were wasps. Researchers in Arizona painted onto "blue-collar" female wasps the dots typical of high status female wasps. The blue collars got beaten up by all the other female wasps.[12]

Notes

[1] Sinkkonen, A. (2009). Umbilicus as a fitness signal in humans. *FASEB Journal* 23 (1), 10-2.

[2] Wise, P. M., Eades J., Tjoa S,, Fennessey, P.V., Preti, G. (2011). Individuals reporting idiopathic malodor production: demographics and incidence of trimethylaminuria. *Am J Me*d 124 (11), 1058-63

[3] Sorokowski, P. (2008). Attractiveness of blonde women in evolutionary perspective: studies with two Polish samples. *Perceptual and Motor Skills 106 (3), 737-744.*

[4] Guéguen, N. (2012). The sweet color of an implicit request: women's hair color and spontaneous helping behavior. *Social Behavior and Personality* 40 (7), 1099-1102(4).

[5] Lynn, M. (2008). Determinants and consequences of remale attractiveness and sexiness: realistic tests with

restaurant waitresses. *Archives of Sexual Behavior* 38 (5), 737-745.

[6] Mannes, A. E. (2012). Shorn Scalps and Perceptions of Male Dominance. *Social Psychological and Personality Science* 4 (2), 198-205.

[7] Burris, R. P., Rowland, H. M., Little, A. C. (2009). Facial scarring enhances men's attractiveness for short-term relationsh. *Personality and Individual Differences* 46 (2), 213-217.

[8] Herbenick D., Schick V., Reece M., Sanders S., Fortenberry J.D. (2010). Pubic hair removal among women in the United States: prevalence, methods, and characteristics. *Journal of Sexual Medicine* 7 (10), 3322-30.

[9] Fahs, B. (2013). Shaving it all off: Examining social norms of body hair among college men in a Women's Studies course. *Womens Studies* 42, 559-577.

[10] Barber, N. (2001). Mustache fashion covaries with a good marriage market for women. *Journal of Nonverbal Behavior* 25 (4), 261-272.

[11] Mansell, K. (2014). Recreating a 2,000-year-old

cosmetic. *Nature.* doi:10.1038/news041101-8.

[12] Strassmann, J. E. (2004). Animal behaviour: Rank crime and punishment. *Nature.* doi:10.1038/432160b

Some Like It Too Hot

Imagine a medical advisory discreetly mailed to unfaithful men everywhere. "Warning," it says. "Extramarital sex can kill."

The medical staff of the Andrology Clinic at the University of Florence has never distributed any such advisory. But maybe it should. In an *International Society of Sexual Medicine* review of the literature on infidelity, members of the clinic's staff presented intriguing evidence that sudden coital death in men is largely the problem of adulterers.

Since the 1970s, physicians have known that, for most men, sex is safe, and even life-prolonging. But in 1963 a Japanese pathologist reported that, of 34 men who had died during intercourse, most had died of cardiac causes, and nearly 80% had died during extramarital sex. In 2005 Korean pathologists documented 14 cases of sudden coital death and found that all had died of cardiovascular causes, and only one was consorting with the woman known to be his wife. In 2006 researchers from the University of Frankfurt published an analysis of sex-related autopsy reports for 68 men. Ten had died with a mistress, and 39 with a prostitute. Only 40 of the autopsies included significant medical history, but almost all of those indicated strong cardiovascular risks. And the University of Florence team's own 2011 statistical analysis of health outcomes for almost 1,700 male patients seen at their clinic showed that being unfaithful represents an independent risk factor for cardiovascular emergencies.

"We were surprised," says Alessandra Fisher, the study's lead author, "especially given the fact that we had recently documented unfaithful men in our patient population as having, in general, better vascular flow, larger testicles, and higher levels of male hormones."

If unfaithful men are hale and hearty, why do they die doing what they love to do?

One clue may lie in a 2011 University of Maryland School of Medicine study of blunt trauma injuries to the penis. Penile fractures, it said, usually result from sex in "unusual social situations" like an elevator or public restroom. More than half of the 16 patients who required reparative surgery had been injured during extramarital affairs.

"Extramarital sex may have its own hazards," says Fisher. "For example, the lover might be much younger. Sex might be particularly athletic or follow excessive drinking or eating."

Does secrecy also add to the cardiovascular load? Can guilt cause fatal physiological stress? Is there something moderating about sex with someone who really cares about you? Is that necessary for men with less than brilliant hearts?

Fisher says that more studies are needed.

For More Information

Fisher, A. D., Bandini, E., Rastrelli, G., Corona, G., Monami, M., Mannucci, E., and Maggi, M. (2012). Sexual and cardiovascular correlates of male unfaithfulness, *The Journal of Sexual Medicine*. doi: 10.1111/j.1743-6109.2012.02722.x.

SCIENCE AND LUST

Why Some Men Cheat

According to a University of Connecticut study, there may be risks to a marriage when a woman is the primary breadwinner. Specifically, husbands who make less money than their wives appear to cheat more. Is this a kick in the pants for hard-working women, or what?

In "Her Support, His Support: Money, Masculinity, and Marital Infidelity," published in the May 2015 *American Sociological Review*, sociologist Christin Munsch suggests that young men who are economically dependent on their wives are more likely to have extramarital affairs than those who aren't. Munsch reviewed data from the National Longitudinal Survey of Youth collected between 2001 and 2011 on married people ranging in age from 18 to 32 years. The data showed that, on the whole, approximately 15% of the men who were completely financially dependent on their wives had affairs. That's roughly 3% more than the average married man in the same data set. While it's not an earth-shattering difference statistically, couple-by-couple it can certainly be marriage-shattering.

Munsch's paper does not draw a bold arrow linking cause to effect. Still, the association between infidelity and economic disparity is pronounced enough that, in a press release issued by the American Sociological Association, she speculated:

“Extramarital sex allows men undergoing a masculinity threat—that is not being primary breadwinners, as is culturally expected—to engage in behavior culturally associated with masculinity.... For men, especially young men, the dominant definition of masculinity is scripted in terms of sexual virility and conquest, particularly with respect to multiple sex partners. Thus, engaging in infidelity may be a way of reestablishing threatened masculinity. Simultaneously, infidelity allows threatened men to distance themselves from, and perhaps punish, their higher earning spouses.”

In other words, feeling feminized by their inability to pull in a larger income, some husbands may try to prove themselves "virile" by sexual conquest.

Ironically, high-earning wives in this survey cheated less often than low-earning ones. According to Munsch's data analysis, about 9% of the young wives had affairs overall, but among wives who earned far more “bread” than their husbands, the prevalence of infidelity was only about 5%.

One Country's Porn is Another's Pablum

"All sex is rape." Remember that famous quip by feminist Catherine McKinnon? It got everybody's hackles up—even lots of women's.

Well, according to the website Snopes.com, McKinnon never said it. *Playboy* attributed the quote to her in an attempt to undermine her credibility. But McKinnon did attack pornography as male-centric and dehumanizing, and so did Andrea Dworkin and other feminist scholars of the 1960's. And, really, who could argue with them? It was.

But porn isn't that way any more—at least not in Norway. Well, at least not in all of the ways all of the time. In Norway.

Not according to three researchers from the University of Hawai'i at Manoa, who trained a team of assistants to evaluate porn images from Norway, the United States, and Japan. They chose the three countries because of the disparate ratings they had received on a measure developed by the U.N. to assess gender equality. Of 93 countries evaluated, Norway had been ranked #1, indicating that women there face relatively few disadvantages. By comparison, the United States received a rank of only #15. Japan had been ranked a lowly #54.

As research assistants sat down to page through the pornographic images collected from popular magazines, their bosses predicted that countries ranking higher on the gender equality measure would prove to have produced pornography showing empowered women. Those ranking lower on the measure would have produced pornography of—well, of the other kind.

The researchers had developed a 21-point scale so that the assistants could objectively code "yes" and "no" answers to questions asking, for example, whether the woman pictured was wearing a dog collar, whether she was restrained, whether she was overly young, whether the image focused exclusively on her genitals, whether she was positioned in a way that might provide her pleasure, and whether she had a surgically enhanced body. Hour by hour by hour by hour the assistants rated porn....

... And the envelope please....

In January 2012 the *International Journal of Cultural Relations* published "Are Variations in Gender Equality Evident in Pornography? A Cross Cultural Study." As expected, the assistants had scored Norwegian pornography as the most egalitarian, featuring a variety of body types, ages, and physical positions, and in general showing everyday women actually enjoying themselves. U.S. and Japanese porn, on the other hand, generally portrayed young, ideal bodies posed in ways that couldn't be expected to be comfortable, much less fun for the women.

But apparently something is still rotten in Scandinavia. For as egalitarian as the assistants found Norwegian porn to be, they still judged it to be "demeaning" to the women portrayed.

Ah, but maybe not. The report concluded by acknowledging that "demeaning" was a subjective judgment, whereas the yes/no answers on the 21-point scale had been objective. Worse, before allowing their assistants to make that judgment, the researchers had not screened them for their biases about porn or sex.

One other big error: The researchers didn't factor into their statistical calculus the impact of a country's porn laws on the porn it produced. This means that Japan's scores, for example, might be out of whack. There it is a crime to picture a vagina—and that renders meaningless any survey question about whether a Japanese picture focuses exclusively on a woman's genitalia.

Even given these drawbacks, the study seems to have produced evidence that at least one country with relative gender equality produces porn that is somewhat egalitarian. The really good news may be that, as the status of women rises country by country, women may find porn fun and enlivening. It may even put a twinkle back in the eyes of the sex-weary. The runaway success of the novel *Fifty Shades of Grey* (and its sequels), which tells a racy story from the point of view of a likable young woman with awakening desire, certainly indicates that there's a market for the "I'm OK with this if you're OK with this" stuff.

But come to think of it, *Fifty Shades of Grey* doesn't have pictures. Any pictures.

Way back in 2008, with the election of Barack Obama to the presidency, we at last began to welcome the idea of a government that looks like America. I like to think that we're egalitarian enough to be ready for porn stars who do. But for some reason I'm not yet convinced.

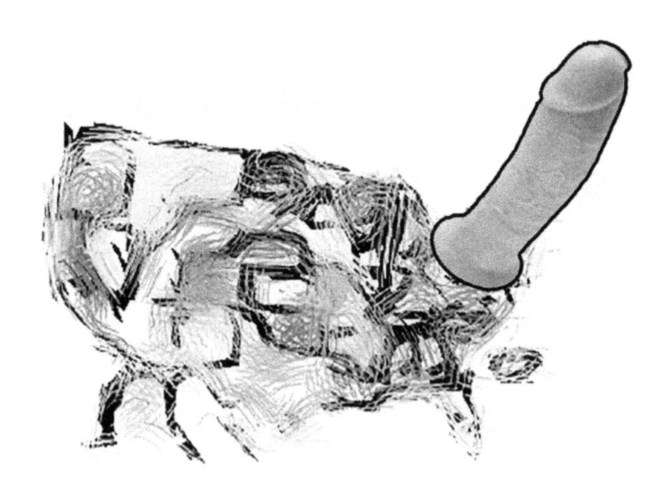

Was It Good For You, Too?

You can ask that question, but if you're a man you might not want to trust the answer.

Between 1969 and 2007, 132 studies were published comparing self-reports on sexual arousal with actual physiological measurements. More than four thousand people (2,505 women and 1,918 men) participated in one or more of these studies.

Then, in 2007, a team from Canada and The Netherlands systematically reviewed the statistical data from all the studies. They found that 66% of men's self-reports accurately reflected the physiological changes that sexual fantasy and erotica provoked in their genitalia and throughout their bodies. But for only 26% of women did the self-reported measures of pleasure and the actual bodily changes line up.

Why?

The researchers suggested many factors that might make

women either underreport or misjudge the extent of their arousal. For example, a woman's menstrual cycle might throw off her sensitivity to stimuli. The hormones in a birth control pill might dim or enhance desire. Women don't have penises that stiffen and grow and become hard to ignore. So, relative to a man, the average woman might be dim-sighted about what's happening for her genitally. There may be cognitive differences between men and women, making women less likely than men to remember sexual arousal long enough to report about it.

And maybe science does not yet not know how to accurately measure the pleasure responses of women. What, for example, do audio cues like sighs signify in a woman as opposed to a man? Should vaginal pulses be counted and measured when reasonable people disagree about whether the vagina is a pleasure center? And so on.

While the researchers raised questions such as these, they didn't definitively answer any of them. Instead, they used their questions to propose parameters for the design of new sex studies.

Clearly, for studies of women's sexual response, both self-reports and physiological measures should be taken into consideration, as they disagree far more often than they agree.

The researchers made one suggestion for studies of men that, while it seems practical, also smacks of a *Clockwork Orange* world. Self-reports of men and physiological data from men agree 66% of the time. So, in a time of tight budgets, can studies dispense altogether with electronic monitoring and base conclusions on self-reports alone?

The researchers gently suggested that the answer is "No." There are men who would purposefully lie about sexual proclivities. For example, from habitually brutal men or men inclined to rape or pedophilia, honesty about excitement and desire should probably not be assumed.

And so (the researchers suggested), in studies of men's arousal and desire, electrodes on the head, the heart, and the penis might be a good idea.

And Why Not?

While the question of why women are not as able as men to assess their own sexual arousal and desire remains unanswered, the researchers did address a related conundrum: Why haven't millions of years of evolution within the human line relieved women of their "disability?"

The answer is that it may be an ability in disguise.

If Charles Darwin was right, like all animals, humans have a biological imperative to procreate and then ensure the survival to adulthood of their progeny.

In order to engage in the procreative act, men need both to feel aroused and to know it.

Women? Not so much.

Desire has nothing to do with getting pregnant. And being ruled by it may be counter to the interests of children's survival.

Indeed, by remaining monogamously devoted to her children's father regardless of her sexual arousal at the sight or sound of anyone else, a woman might resolve any questions her mate has about his children's paternity. The woman's ongoing fealty may also encourage the father to provide his children with sustenance and protection long-term.

If so, heightened sensitivity to his own arousal helps a man propagate the species. But for a woman, the opposite may be true. Dimming her desire—being excessively choosy about mating, and basing her choice not on a man's sexual prowess but on his ability to protect and provide—may be just what the biological imperative ordered.

For More Information

Chivers, M.L., Seto, M.C., Lalumière, M.L., Laan, E, Grimbos, T. (2010). Agreement of self-reported and genital measures of dexual arousal in men and women: A meta-analysis. *Archives of Sexual Behavior* 39:5–56 A

Lalumière, M. L., Quinsey, V. L., Harris, G. T., Rice, M. E., and Trautrimas, C. (2003). Are rapists differentially aroused by coercive sex in phallometric assessments? *Annals of the New York Academy of Sciences*, 989, 211–224.

Morokoff, P. J. (1985). Effects of sex guilt, repression, sexual "arousability", and sexual experience on female sexual arousal during erotica and fantasy. *Journal of Personality and Social Psychology* 49, 177–187.

Price, M. E, Pound, N., and Scott, I.M. (2014). Female economic dependence and the morality of promiscuity. *Archives of Sexual Behavior* 43, 1289–1301.

van Lunsen, R., & Laan, E. (2004). Genital vascular responsiveness and sexual feelings in midlife women: Psychophysiologic, brain, and genital imaging studies. *Menopause,* 11, 741–748.

ABOUT THE AUTHOR

Rebecca Coffey is an award-winning science journalist and television documentarian. She contributes online and offline to national magazines such as *Scientific American, Discover*, and *Psychology Today* and to major market newspapers. Coffey is also a novelist and a humorist.

Other books by Rebecca Coffey:

—The Beck & Branch *Brainy Sex* series continues. Look for news of just-released volumes at www.BrainySex.com.

—*Hysterical: Anna Freud's Story* (She Writes Press, 2014). A fact-based novel. Imagine growing up gay in a household where your world-renowned father calls lesbianism a gateway to mental illness. It is always, he said, caused by the father, and it is usually curable by psychoanalysis. Now imagine that he analyzes you.

—*Nietzsche's Angel Food Cake: And Other Recipes for the*

Intellectually Famished (Beck & Branch, 2013). Humor. When Friedrich Nietzsche made angel food cake, did the angel survive the encounter? When Sigmund Freud handled raw fish, where did his thoughts take him? Exactly what did Dorothy Parker mean by the term "Parker House Rolls?" And how did Ernest Hemingway batter testicles? Literary humor for those who enjoy smashing idols fondly, *Nietzsche's* allows 22 cultural monoliths to share "their" succulent recipes. For those who like cookery to insinuate the hard questions, it offers a funny, surprisingly informative, and entirely whirlwind tour of civilization. Not really a cookbook, it's for lovers of literature, history, art, music, and philosophy, for foodies, and for anyone with a good liberal arts education, no matter how vaguely they remember it.

—*Murders Most Foul: And the School Shooters in Our Midst* (Vook, 2012). Journalism. How can parents, teachers, community members, and even school children help ward off school massacres? In an anecdotal style and with a focus on what the FBI says average people can do to help, *Murders Most Foul* carefully examines the entire history of school massacres and parses long-simmering debates about bullies, gun laws, violent media, and zero tolerance policies.

—*Unspeakable Truths and Happy Endings: Human Cruelty and the New Trauma Therapy* (Sidran,

1998). Journalism. With the electrifying tales of 15 survivors of catastrophic human cruelty at its narrative core, *Unspeakable Truths & Happy Endings* journalistically explores the affects of survivors's stories on compassionate listeners -- a group that includes therapists but that also includes friends, family, and even survivors themselves as they work and re-work the realities of their own experience. Along the way, the book addresses the flip side of compassionate listening; squabbles about victimhood and recovered memory. The author writes that, as thinking and caring inhabitants of a menacing world, we must all learn to hear unspeakable truths. At the same time that we risk accepting the truths about violence and degradation that survivors' memories hold, we must reasonably engage critical thinking when memories of violence and degradation stretch the limits of our credulity. We owe it to survivors to listen compassionately; we owe it to ourselves to listen prudently.

CPSIA information can be obtained
at www.ICGtesting.com
Printed in the USA
BVHW042146090421
604663BV00014B/329